Leaping Learners
Education, LLC

For more information and resources visit us at:
www.leapinglearnersed.com

Photo Credits
Cover, ©seb2583/stock.adobe.com; Fish diver graphics © Alexey Bannykh/stock.adobe.com; Pg2 ©ead72/stock.adobe.com; pg. 4 ©seb2583/stock.adobe.com; pg 7-8 ©kajornyot/stock.adobe.com; pg 9-10 ©nancy/stock.adobe.com; pg 11 © 3drenderings/stock.adobe.com; pg 12 ©andrea Izzotti/stock.adobe.com; pg 15 ©dejvudesigns/stock.adobe.com; pg 17 ©DJ/stock.adobe.com; pg 18 ©anibal/stock.adobe.com

All design by Sean Bulger
All other pictures by Sean Bulger or royalty free from Pixabay.com

Special thanks to our reading level consultant
FRANCINE DEL VECCHIO, Ed.D.
Director of the Literacy Center
Professor of Education
Caldwell University

ISBN
978-1-948569-00-2

Dear Parents and Guardians,

Thank you for purchasing a *Fay Learns About* series book! After teaching students from kindergarten to second grade for over seven years, I became frustrated at the lack of engaging books that my students could read independently. To help my students engage with nonfiction topics, my wife and I decided to write non-fiction books for children. We hope to inspire young children to learn about the natural world.

Here at Leaping Learners Education, LLC, we have three main goals:

1. Spark young readers' curiosity about the natural world
2. Develop critical independent reading skills at an early age
3. Develop reading comprehension skills before and after reading

We hope that your child enjoys learning with Fay. If you or your children are interested in a topic that we have not written about yet, send us an email with your topic and maybe your book will be next!

Thank you,

Sean Bulger Ed.M

www.leapinglearnersed.com

Reading Suggestions:

Before reading this book, encourage your children to do a "picture walk," where they skim through the book and look at the pictures to help them think about what they already know about the topic. Thinking about what they already know helps children get excited about learning more facts and begin reading with some confidence.

Preview any new vocabulary words with your child. Key vocabulary words are found on the last few pages of the book. Have your children use each new phrase in their own words to see if they understand the definition.

After previewing the book, encourage your child to read the book independently more than once. After they have read it, ask them specific questions related to the information in the book. Encourage them to go back and reread the relevant section in the book to retrieve the answer in case they forgot the facts.

Finally, see if your child can complete the reading comprehension exercises at the end of the book to strengthen their understanding of the topic!

This book is best for ages 6-8
but. . .
Please be mindful that reading levels are a guide and vary depending on a child's skills and needs.

Fay Learns About . . . Whales

Written by Sean and Anicia Bulger

Table of Contents

Hi! My name is Fay. I love to discover and learn new things. In this book, we will learn about whales. Let's go!

Introduction

Whales are animals that live in all the world's oceans.

There are more than eighty kinds of whales. Some whales are small and some whales are large!

Whales

Even though whales live in the ocean, they are **mammals** like you!

Whales are mammals because they breathe air, have hair, give live birth, and their babies drink their mother's milk.

Whales are not fish.

*Fish breathe water

*Fish have scales

*Fish lay eggs

*Fish have gills

Whales are not all the same. There are two main groups of whales. They are **baleen** whales and toothed whales.

Toothed whales and baleen whales eat different foods and have different bodies.

Baleen Whales

Baleen whales do not have teeth.

Instead they have baleen. Baleen is like a brush.

Baleen

Baleen Whale's Diet

Baleen whales eat by scooping up large amounts of water and food. Then they push out the water and trap the food in their baleen.

Baleen whales eat small fish and krill. **Krill** are small shrimp-like animals.

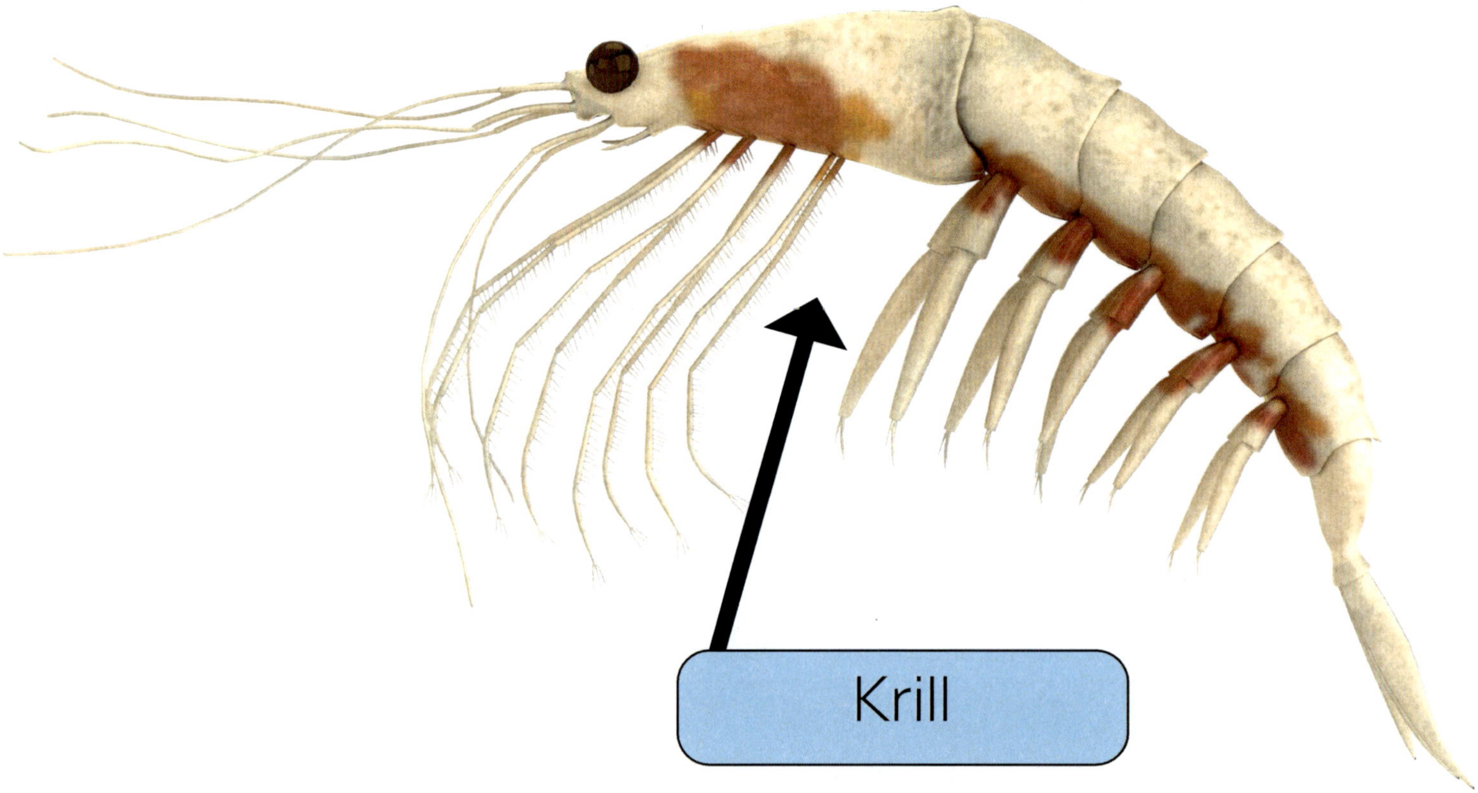

Krill live in large groups near the ocean's surface. This allows baleen whales to scoop hundreds up in one gulp!

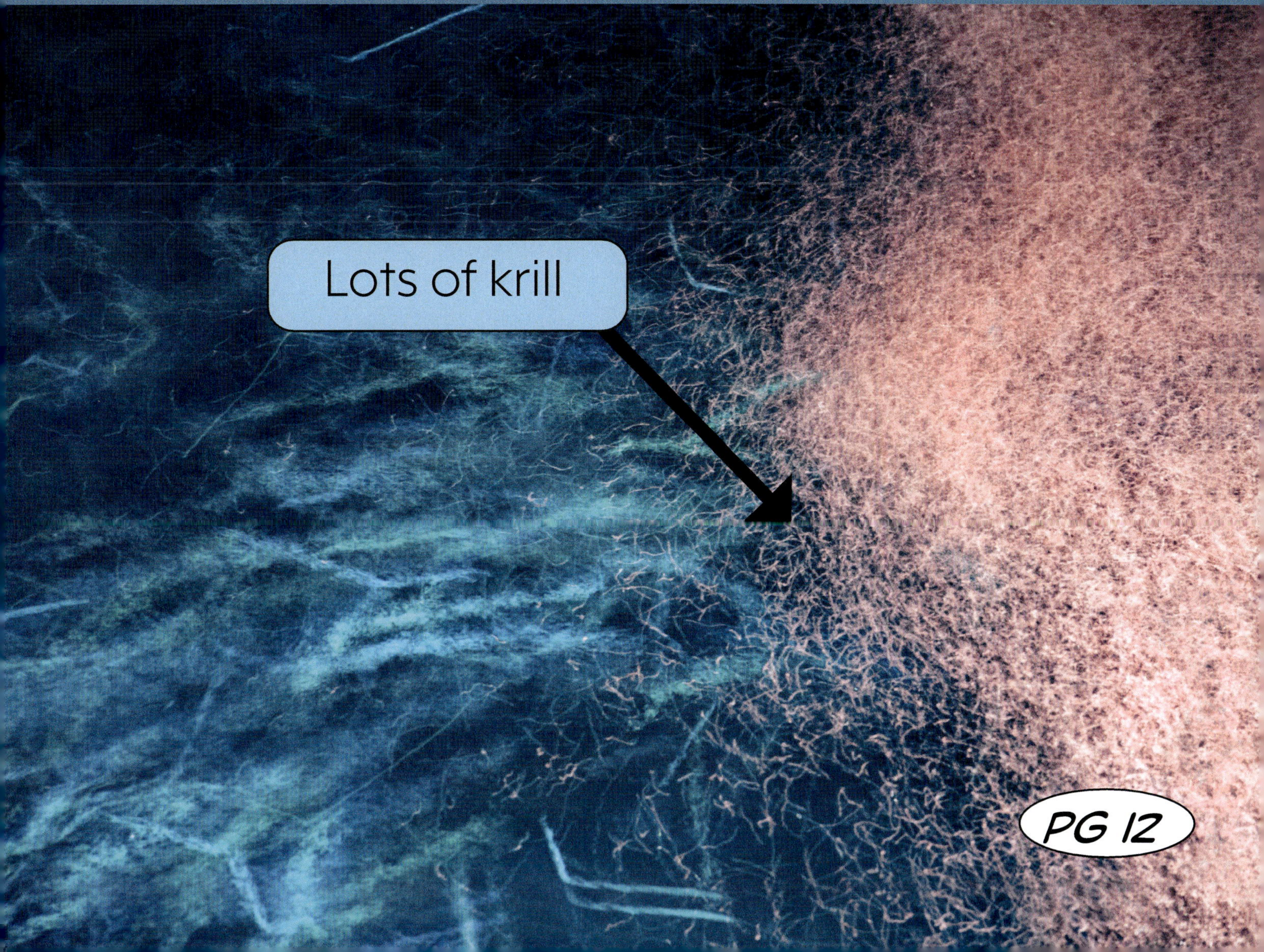

Baleen Whale Body

Baleen whales have two eyes, two ears and two blow holes. They have a **dorsal fin** on top and a **fluke** in back. They have two flippers on their sides.

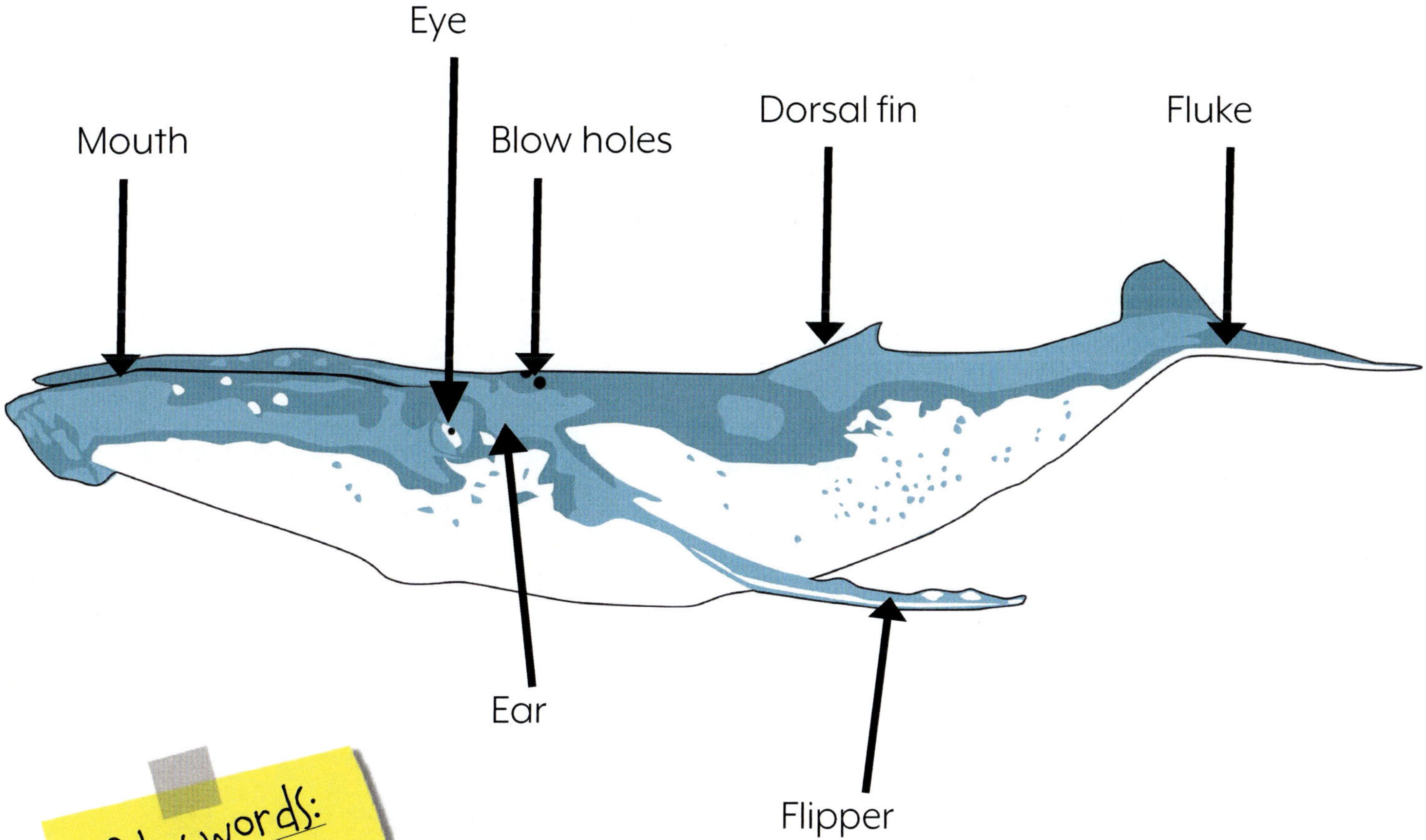
Eye
Mouth
Blow holes
Dorsal fin
Fluke
Ear
Flipper

Tricky words:
chunk it out
dor-sal
Fl-uke

Toothed Whales

Toothed whales look different than baleen whales.

Toothed whales have teeth which they use to hunt and eat their prey.

Toothes Whale's Diet

Toothed whales, like the Orca, hunt seals, fish, and penguins. Other toothed whales eat **squid** and fish.

Fish

Squid

Seal
What do toothed whales eat?

Toothed Whale's Body

Toothed whales have two eyes, two ears and one blow hole. They have a dorsal fin on top and a fluke in back. They have two flippers on their sides.

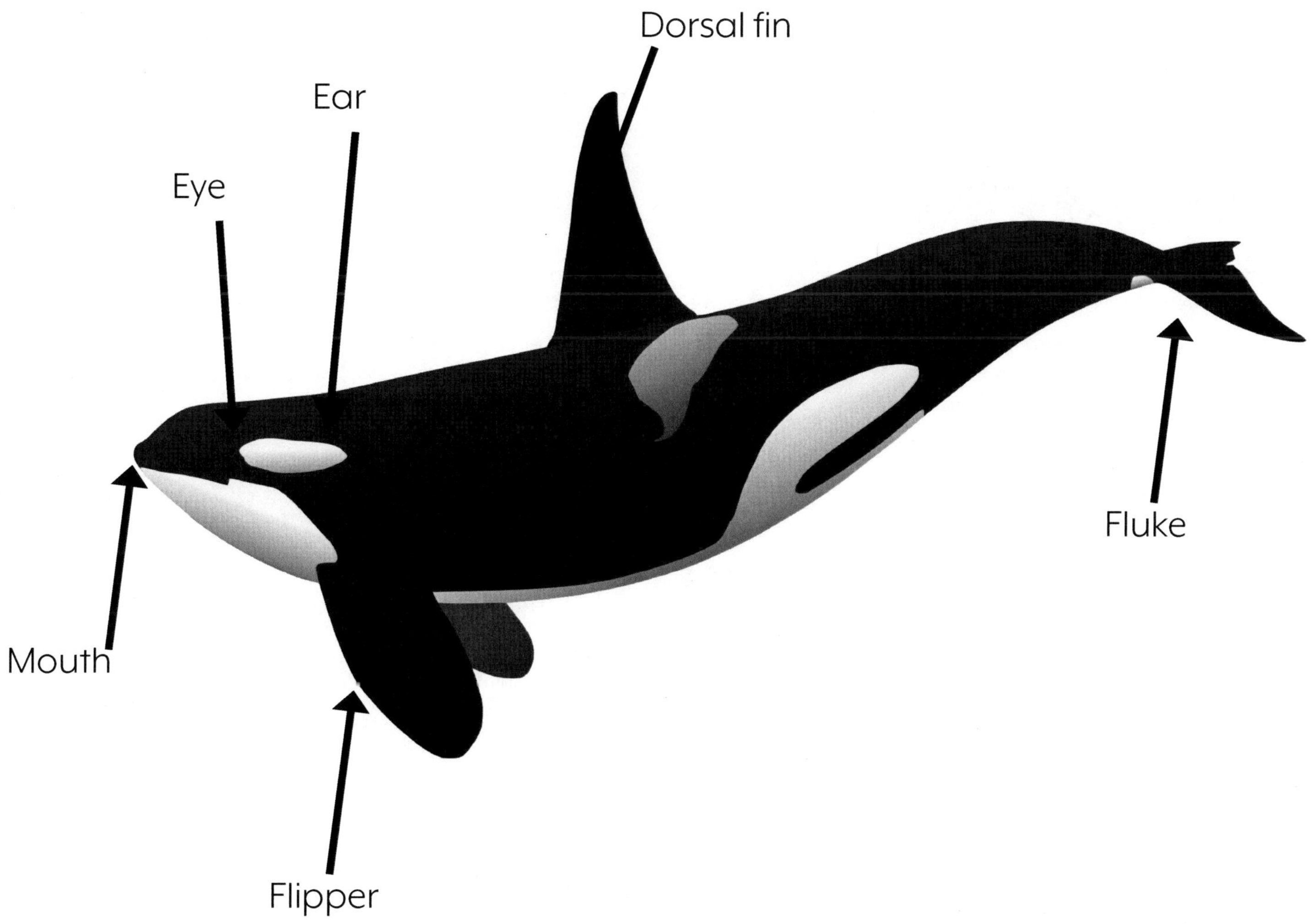
Dorsal fin
Ear
Eye
Fluke
Mouth
Flipper

Breathing

Since whales breathe air, they have a blow hole and lungs.

They use the blow hole to breathe air in and out. Baleen whales have two blow holes and toothed whales only have one.

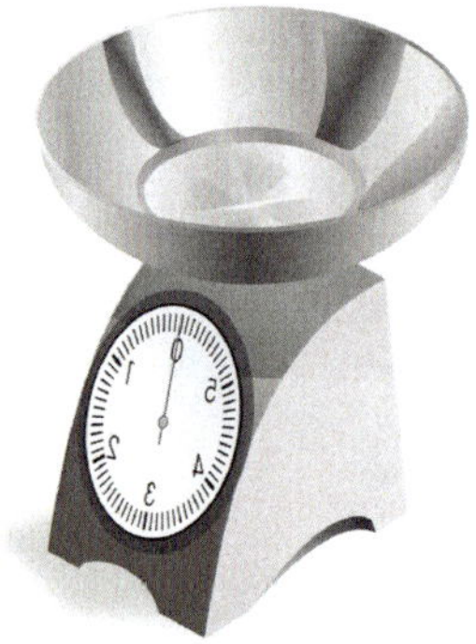

A sperm whale can hold its breath for 1.5 hours!

Dolphins talk with each other using clicks and whistles.

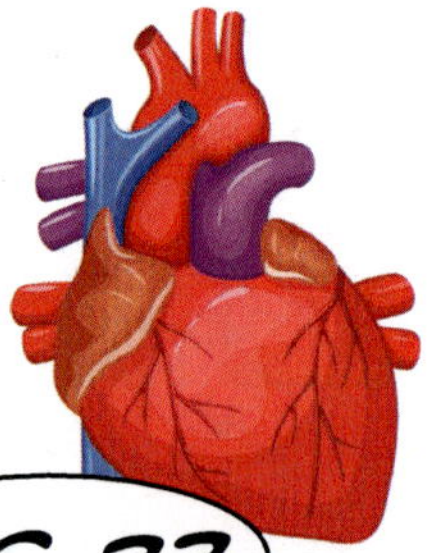

A blue whale's heart is the size of a car!

Glossary

A glossary tells the reader the meaning of important words.

Mammal – A warm blooded animal like people. Mammals give live birth, have hair, breath air and their babies drink milk.

Baleen – A brush like part of a whale's mouth that traps food

Krill – Small shrimp-like animal that baleen whales eat

Dorsal Fin – Top fin on a whale's body

Fluke – Back fin of a whale, similar to a tail

Squid – A long, fast swimming ocean animal with 10 arms

Draw a picture of a whale.

Choose 3 words from the glossary and write a sentence for each one.

1. ______________________________

____

____

2. ______________________________

____

____

3. ______________________________

____

____

Quiz

1. What type of animals are whales?
a. Fish
b. Mammals
c. Reptiles

2. What do baleen whales eat?
a. Squid
b. Krill
c. Plankton

3. What do toothed whales eat?
a. Squid
b. Krill
c. Plankton

Common core standards:
RI. 2. 1 - Questions 1, 2, 3

4. How many blow holes does a toothed whale have?

a. 1

b. 2

c. 3

5. In which section do you learn about the fluke?

a. Body

b. Diet

c. Breathing

6. How long can a sperm whale hold its breath?

a. 1 min

b. 1.5 hours

c. 12 hours

Common core standards:
RI. 2. 1 - Questions 4,6
RI. 2. 2 - Question 5

What is the difference between toothed whales and baleen whales?

Use 1-2 pieces of evidence from the text and answer in complete sentences.

Label the Killer Whale's Body

Want to learn about rainforest animals? Check out the "Matt Learns About..." series!

Great for emerging readers ages 6-8

Want to learn about colors? Check out the "Clayton Teaches You About..." Series!

Great for early readers ages 4-6

Want to learn about Farm Animals? Check out the "Katie Teaches You About..." Series!

Great for early readers ages 4-6

Want to learn about colors? Check out the "Clayton Teaches you About..." Series!

Great for early readers ages 4-6

Want to learn about Farm Animals? Check out the "Katie Teaches you About..." series!

Great for early readers ages 4-6